Food Innovation and Engineering 2019

Topical Conference at the 2019 AIChE Annual Meeting

Orlando, Florida, USA
10 – 15 November 2019

ISBN: 978-1-7138-0559-5

Printed from e-media with permission by:

Curran Associates, Inc.
57 Morehouse Lane
Red Hook, NY 12571

AIChE
120 Wall Street, FL 23
New York, NY 10005-4020

Phone: (800) 242-4363
Fax: (203) 775-5177

www.aiche.org

Additional copies of this publication are available from:

Curran Associates, Inc.
57 Morehouse Lane
Red Hook, NY 12571 USA
Phone: 845-758-0400
Fax: 845-758-2633
Email: curran@proceedings.com
Web: www.proceedings.com

TABLE OF CONTENTS

576029 Bioconversion of Whey Lactose to Acetate and Propylene Glycol Deicers

Whey from cheese and yoghurt production operations contain significant amounts of lactose that is currently being unutilized and disposed of as waste. The treatment and disposal costs for these wastes are high due to the high biochemical oxygen demand (BOD) of these wastes. A novel low pH fermentation process will be discussed in this paper for the bioconversion of whey lactose to acetate and propylene glycol deicers as products. Acetate deicers are biodegradable and less harmful to the environment than the currently used non-biodegradable chloride deicers. A two-stage fermentation process using *Lactobacillus plantarum* in the first stage to convert lactose to lactate, and *Lactobacillus buchneri* in the second stage to convert lactate to acetate and propylene glycol was studied. Final product concentrations of 25 to 30 g/L of acetate and propylene glycol were obtained with lactose as the substrate. The fermentation pH was ~ 4.2, providing for efficient acid recovery. Kinetic data for separate sequential and coculture fermentation data will be presented.

575165 Process Design, Techno-Economic Assessment and Environmental Impacts of Succinic Acid Production from the Organic Fraction of Municipal Solid Waste

This work focuses on process design and techno-economic evaluation of bio-based succinic acid production using the organic fraction of municipal solid waste (OFMSW) as crude feedstock. The upstream, fermentation and downstream separation and purification stages were simulated using the process simulator UniSim (Honeywell). The OFMSW contains free sugars, starch, cellulose, proteins and oils/fats. The OFMSW was initially fractionated for the separation of oils/fats and proteins followed by enzymatic hydrolysis of polysaccharides. Succinic acid production was carried out by fed-batch fermentations using either the bacterial strain *Actinobacillus succinogenes* or the genetically engineered yeast strain *Yarrowia lipolytica*. The optimal fermentation efficiencies (*i.e.* final succinic acid concentration, yield and productivity) were used. The recovery of succinic acid crystals was based on bacterial cell removal, decolourisation of fermentation broth using activated carbon, acidification of organic salts by ion-exchange resins, evaporation and drying. Techno-economic evaluation was carried out at various plant capacities. The greenhouse gas emissions and the non-renewable energy requirements were estimated.

Acknowledgements

Financial support has been provided by the STAR-ProBio project funded by the European Union's Horizon 2020 Research and Innovation Programme under Grant Agreement No. 727740

573759 Bioprocess Design and Techno-Economic Evaluation of 2,3-Butanediol Production from Crude Glycerol and Sugar-Based Feedstocks

2,3-Butanediol (BDO) production from crude renewable resources is a potential platform chemical for the development of a sustainable chemical industry. Fed-batch fermentations for BDO production have been optimised using very high polarity sugar and molasses derived from sugarcane mills as well as crude glycerol from biodiesel production plants. Fermentation media, volumetric oxygen transfer coefficient ($k_L a$), temperature and pH were considered in the optimization of individual fed-batch cultures. The optimal fermentation efficiencies (i.e. fermentation medium composition, yield, productivity, final BDO concentration) were used in bioprocess design simulations using the software UniSim (Honeywell) that was used to validate material and energy balances. The downstream separation and purification section was based on the conversion of BDO into an acetal that is immiscible with water followed by distillation steps. An alternative downstream separation and purification process was also evaluated based on the direct conversion of BDO into methyl ethyl ketone (MEK). Preliminary techno-economic evaluations were carried out for all feedstocks, associated fermentation efficiencies and downstream separation alternative process flow sheets showing the bottlenecks for industrial implementation. Sensitivity analysis was subsequently conducted considering the most important parameters in order to identify the process improvements required for the industrial implementation of BDO or MEK production. The results presented in this study demonstrate that crude glycerol is a promising feedstock for the production of BDO and MEK.

Acknowledgements

The work presented in this study has been funded by the National Agency of Petroleum (ANP), Petrobras, Brazil (project 00320-2/ 2012

577105 Design and Analysis of the Integrated Process for Algae Conversion to Mix Alcohols

Design and Analysis of the Integrated Process for Algae Conversion to mix Alcohols

Yue Zhang, Abhay Athaley, Marianthi Ierapetritou

Department of Chemical and Biochemical Engineering, Rutgers - The State University of New Jersey

Petroleum is the main raw material for the production of most commonly used fuels and chemicals. However, due to exhausting resources and emerging environmental issues from oil industry, alternative production paths have been investigated. Biomass sources are among the alternatives with the potential to produce both chemicals and fuels[1]. Particularly, algae has attracted a lot of interest as a third generation biomass that contains no lignin, and requires no land or fertilizer [2].

The acceptance of bio-products in the market closely relates to the competitiveness with petroleum-based products not only in terms of economics but also in terms of process sustainability. A large number of researches have focused on the biomass conversion to biofuel[3, 4] and the MixAlco® process is a widely studied bio-refining technology which can achieve this goal. The MixAlco® process can convert any biodegradable material into mixed alcohol fuels [5-7]. The ketonization route and the esterification route are the two main routes for MixAlco® process, and their economic, energy, and carbon footprint assessments have been discussed[8]. In our work, we investigate and integrate the novel fermentation process with the two downstream process to produce higher carbon alcohols with better efficiency. A synthetic syntrophic consortia of Clostridium kluyveri and Clostridium ljungdahlii is applied to the fermentation process[9-11]. The co-existence of these two microbes can enable fast and efficient utilization of carbon and electrons in the substrates by importing additional electrons as needed from H_2.

This work uses techno-economic analysis and life cycle assessment to design and evaluate the production alternatives for producing mix alcohols. The detailed process flowsheet is developed and simulated using Aspen Plus. Four different routes are simulated to produce alcohol mixture: 1) Ketonization Route; 2) Ketonization with Novel Fermentation Route; 3) Esterification Route; and 4) Esterification with Novel Fermentation Route. Heat integration and economic analysis is carried out to calculate the minimum product selling price of the alcohols for each route. Sensitivity analysis is performed to analyze the bottlenecks of each route and to compare the cost of the products with fluctuating cost of algae and capacity of the plant. To check the competitiveness of the above routes in terms of sustainability, Life Cycle Analysis is carried out using SimaPro®. The results with respect to carbon production and water consumption are discussed and compared to the current oil-based processes.

References:

1. Bozell, J.J., *Feedstocks for the future–biorefinery production of chemicals from renewable carbon.* CLEAN–Soil, Air, Water, 2008. **36**(8): p. 641-647.
2. Jung, K.A., et al., *Potentials of macroalgae as feedstocks for biorefinery.* Bioresource Technology, 2013. **135**: p. 182-190.
3. Marchetti, J., V. Miguel, and A. Errazu, *Techno-economic study of different alternatives for biodiesel production.* Fuel Processing Technology, 2008. **89**(8): p. 740-748.
4. Zhang, Y.-H.P., *What is vital (and not vital) to advance economically-competitive biofuels production.* Process Biochemistry, 2011. **46**(11): p. 2091-2110.
5. Holtzapple, M.T., et al., *Biomass conversion to mixed alcohol fuels using the MixAlco process.* Applied Biochemistry and Biotechnology, 1999. **77-9**: p. 609-631.
6. Granda, C.B., et al., *Carboxylate Platform: The MixAlco Process Part 2: Process Economics.* Applied Biochemistry and Biotechnology, 2009. **156**(1-3): p. 537-554.
7. Pham, V., M. Holtzapple, and M. El-Halwagi, *Techno-economic analysis of biomass to fuel*

conversion via the MixAlco process. Journal of Industrial Microbiology & Biotechnology, 2010. **37**(11): p. 1157-1168.
8. Fasahati, P. and J.J. Liu, *Application of MixAlco (R) processes for mixed alcohol production from brown algae: Economic, energy, and carbon footprint assessments*. Fuel Processing Technology, 2016. **144**: p. 262-273.
9. Yin, Y.N., et al., *Biological caproate production by Clostridium kluyveri from ethanol and acetate as carbon sources*. Bioresource Technology, 2017. **241**: p. 638-644.
10. Richter, H., et al., *A Narrow pH Range Supports Butanol, Hexanol, and Octanol Production from Syngas in a Continuous Co-culture of Clostridium ljungdahlii and Clostridium kluyveri with In-Line Product Extraction*. Frontiers in Microbiology, 2016. **7**: p. 13.
11. Ragsdale, S.W. and E. Pierce, *Acetogenesis and the Wood–Ljungdahl pathway of CO2 fixation*. Biochimica et Biophysica Acta (BBA)-Proteins and Proteomics, 2008. **1784**(12): p. 1873-1898.

579325 Production of Value Added Co-Products from the Hydrothermal Liquefaction of Unhydrolyzed Solids

Catalytically assisted hydrothermal treatment of corn stover was performed in a flow through reactor at 170-190 °C to separate lignin and hemicellulose from the biomass. The sugars obtained from such pretreatment process can be used for bio-ethanol production. The liquid hydrolysate obtained under optimized operating conditions, consists of partially solubilized lignin components and hemicellulose hydrolysate as co-products. This aqueous liquor was mixed with unhydrolyzed solids (UHS) obtained from the pre-treatment and enzymatic hydrolysis of corn stover and hydrothermally treated at 300°C and 1600 psi for 1 hour. Following the HTL reaction, the char obtained from separation of biocrude and bio-oil was thermally treated at 400°C for 2 hours and its specific surface area was measured using BET analyzer. Thermally treated char was used as electrode material for supercapacitors and its capacitance was measured using cyclic voltammetry at the scan rates of 5 mV/s – 100 mV/s. The biocrude obtained was further analyzed for lactic acid and phenols using HPLC and GCMS respectively Results obtained for the pre-treatment for corn-stover, HTL reaction, and characterization will be presented.

569933 Hydroxyl Group Stabilization for Increased Yields of Low Molecular Weight Products in the Co-Pyrolysis of Cellulose and Thermoplastics

Plant-derived biomass as a renewable resource for value added chemical production has the potential to meet the needs of reducing the usage of fossil fuels even with growing global population. Fast pyrolysis is one of the most simple and robust thermochemical methods for biomass conversion, but producing value added chemicals via fast pyrolysis is still challenging due to diverse production distributions, potentially incurring significant cost due to subsequent upgrading and separation operations. For inexpensive and sustainable chemical production processes, it is crucial to understand reaction mechanism of the pyrolysis of cellulose - the largest fraction of biomass. Cellulose pyrolysis typically produces anhydrosugars, including levoglucosan (LG) and its oligomers, and low molecular weight products (LMWPs), which can be potentially upgraded into biofuels or value-added chemicals. In this work, we present a strategy to promote the yields of LMWPs from cellulose pyrolysis via hydroxyl group stabilization using molten polymers. Three types of thermoplastics, high-density polyethylene (HDPE), polyethylene glycol (PEG), and polystyrene (PS), representing different functional groups (e.g., aromatics, ethers, etc.), are individually co-pyrolyzed with cellulose to investigate the roles of these moieties in cellulose pyrolysis. The experimental results show that the yields of LMWPs can be significantly increased in the presence of PS, suggesting that the aromatic groups in PS stabilize the cellulosic hydroxyl groups during glycosidic bond cleavage, inhibiting the reactions leading to anhydrosugar molecules. Our experiments also suggest that both ether and aromatic groups stabilize the hydroxyl groups in glycosyl rings during dehydration, leading to increased yields of products from retro Diels-Alder fragmentation and decreased yields of dehydration products. Our work demonstrates that by manipulating the compositions of the molten polymers in cellulose/molten polymer co-pyrolysis, product distributions can be manipulated and yields of desired products can be enhanced.

567679 Additive Manufacturing of Food: Formulation and Printability Criteria

The recent explosion of interest in 3D printing technologies has forced a number of different industries to evaluate its potential as a viable production technique. While great successes and advances are clear in certain industries such as medical devices and aerospace other areas such as the food industry have not yet found wide use for the technology.

The ability to be able to selectively deposit material or molecules within a 3D volume and hence, gradate the composition (e.g. via direct manufacture of pores), offers the possibility for the control of complex structures for altering texture, taste & functionality in food products. Such manipulation of microstructures can allow regulation of fracture, breakdown or dissolution mechanics during product use, opening up the possibility of a range of functional and novel foods.

While this manufacturing technique is largely unheard of in large scale food production there is significant interest and innovation in crude 'macro' scale food printing for the catering industry and niche food product markets. To drive this technology forward a significant increase in the knowledge and design rules of potential food materials for printing is needed coupled with a focus on controlling structures at a smaller scale to build more complex multicomponent systems. It is suggested that with this knowledge significantly more sophisticated, and personalised products can be offered especially when coupled with a point of sale or distributed manufacturing model.

Among the current 3D printing methods for food applications (Chia & Wu, 2015; Guo & Leu, 2013), extrusion is a prevailing technique because it is easy to develop and it has the broadest set of "inks"(Guvendiren et al., 2016; Tan, Toh, Wong, & Lin, 2018). These inks are often divided into three groups depending on the extrusion techniques (Godoi, Prakash, & Bhandari, 2016): cold extrusion, hot-melt extrusion, and gel-forming extrusion. Both hot-melt and gel-forming extrusion must possess gel-forming mechanisms (Kirchmajer, Gorkin III, & in het Panhuis, 2015; Sun, Zhou, Huang, Fuh, & Hong, 2015). In hot-melt extrusion, semisolid ink is extruded at a relatively high temperature from the nozzle and it needs to solidify almost immediately after extrusion to weld to the previous layer (Sun, Zhou, Yan, Huang, & Lin, 2018). If the gel-forming mechanism is based on chemical cross-linking the reagents are often harmful and are unlikely to be used for food design. In case of ionotropic cross-linking and complex coacervate formation (Godoi et al., 2016), the number of edible materials that can be used as inks is limited. However, in cold extrusion the ink is made of a self-supporting material and the extrusion is generally conducted at room temperature (Sun et al., 2018).

This study initially investigates rheological properties and printability (shape fidelity) of food-grade hydrocolloid pastes. The results show that the phase angle and the relaxation exponent could be used to understand the solid and liquid characteristics of paste behaviour during the cold extrusion 3D printing process. If the phase angle is in the range of 3° - 15° and the relaxation exponent is in the range of 0.03 - 0.13 , the ink is self-supporting (Figure 1). Since measuring phase angle in the viscoelastic region is relatively straight forward, the phase angle could be used as a quick and effective means of studying the printability of the formulation. The obtained knowledge can be used as a design rule for food (hydrocolloids) printing processes to develop new feedstocks for food 3D printing. Further progress to expand these design rules into other edible material classes is ongoing.

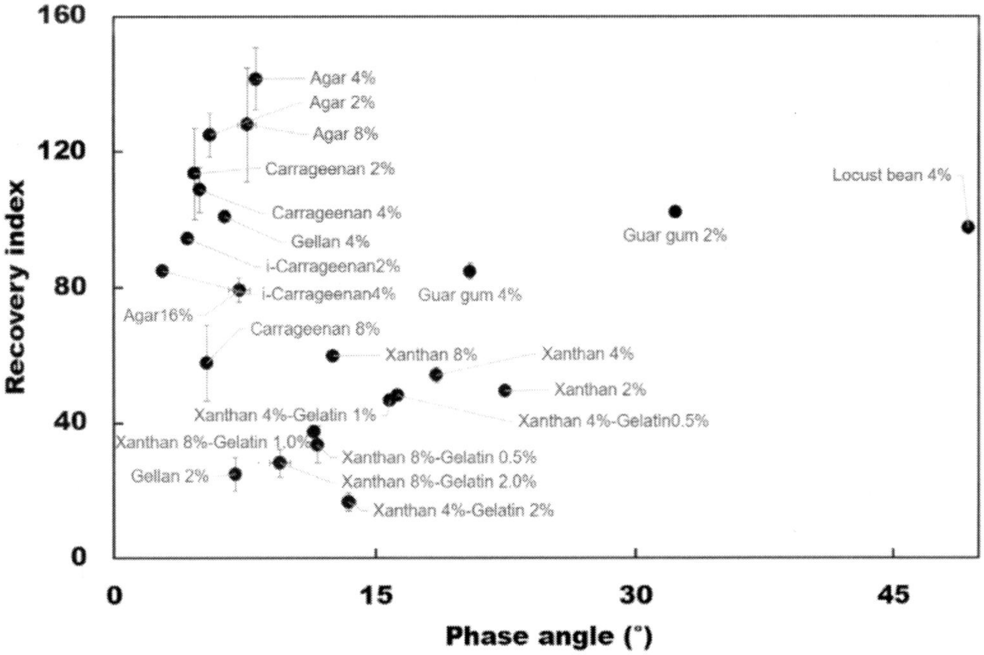

Figure 1 Recovery index vs phase angle for the re-structured samples

Chia, H. N., & Wu, B. M. (2015). Recent advances in 3D printing of biomaterials. *Journal of Biological Engineering*, *9*, 4. https://doi.org/10.1186/s13036-015-0001-4

Godoi, F. C., Prakash, S., & Bhandari, B. R. (2016). 3d printing technologies applied for food design: Status and prospects. *Journal of Food Engineering*, *179*, 44–54. https://doi.org/10.1016/j.jfoodeng.2016.01.025

Guo, N., & Leu, M. C. (2013). Additive manufacturing: Technology, applications and research needs. *Frontiers ofMechanical Engineering*, *8*(3), 215–243. https://doi.org/10.1007/s11465-013-0248-8

Guvendiren, M., Molde, J., Soares, R. M. D., & Kohn, J. (2016). Designing Biomaterials for 3D Printing. *ACS Biomaterials Science and Engineering*, *2*(10), 1679–1693. https://doi.org/10.1021/acsbiomaterials.6b00121

Tan, C., Toh, W. Y., Wong, G., & Lin, L. (2018). Extrusion-based 3D food printing – Materials and machines. *International Journal of Bioprinting*, *4*(2), 1–13. https://doi.org/10.18063/ijb.v4i2.143

Kirchmajer, D. M., Gorkin III, R., & in het Panhuis, M. (2015). An overview of the suitability of hydrogel-forming polymers for extrusion-based 3D-printing. *J. Mater. Chem. B*, *3*(20), 4105–4117. https://doi.org/10.1039/C5TB00393H

Sun, J., Zhou, W., Yan, L., Huang, D., & Lin, L. ya. (2018). Extrusion-based food printing for digitalized food design and nutrition control. *Journal of Food Engineering*, *220*, 1–11. https://doi.org/10.1016/j.jfoodeng.2017.02.028

567684 The Possible Application of a Single Passive Acoustic Emission Sensor to Identify Different Complex Fluids in a Fully Flooded Pipe

Abstract

The ability to measure process and product parameters are key for any manufacturing process. Modern sophisticated manufacturing heavily relies on constant measurements in order to guarantee a consistent product quality. Recent trend is the move towards smart manufacturing or more commonly referred as *Industry 4.0*; a term first introduced by the *Association of German Engineers* and the German government in 2011, which simplified describes the need to make use of networks and big data in order to meet modern manufacturing and processing needs [9].

One example of a technology delivering high volumes of data and information are acoustic emission sensors (AE sensor).

Such sensors can either be *active*, meaning they are based on an emitter-receiver system, or *passive*. Where *active* acoustic emission sensing measures the change of introduced acoustic waves over the distance from the emitter to the receiver, *passive* acoustic emission sensors only detect the acoustic emission that the process emits itself. Although *active* acoustic emission sensors are more present in research and in industry they face several challenges, which include loss of signals due to obstructions, bubbles and distance or simply not reaching the necessary penetration depth [1, 7]. Overall, *active* acoustic emission sensors prove to give good predictions on factors such as flow rate, degree of gassing or solid content. This technology works well for Newtonian and non-Newtonian fluids [11, 15].

There have been several attempts to make use of acoustic emission data to gain a greater understanding of pipe flow [3, 8, 12], however, work has mainly been focussed on active acoustic emission [2, 5], multiphase flow [8, 14] or water [10, 12, 13].

Passive acoustic emission sensors are used for leak detection in water pipes by employing in-pipe hydrophones [4, 10] or by aiming to recognise acoustic patterns based on the signals of a series of sensors [12]. Other work on passive acoustic emission sensing has mainly focused on multiphase systems [6, 8, 14]. However, no work was done so far on enclosed, fully flooded and single-phased pipe systems.

The work presented makes use of a single passive acoustic emission sensor, which is placed on the outer wall of a 1-inch (ID) stainless steel pipe (L=120 mm) together with a new, self-developed device that makes is possible to receive relevant *AE* signals. Along with *AE* signal recording 2D Particle Image Velocimetry (2D PIV) is used in order to visualise the flow fields for different fluids and obstruction types. The 2D PIV chamber used meets the same inner diameter as the stainless steel pipe, hence, is assumes to be representative for the flow fields within it. Four in-pipe obstructions are used on four different test fluids, which all find application in the FMCG sector. The two Newtonian fluids used are water and glycerol (CAS Number: 56-81-5), whereas *Carbopol®* (CAS Number 9003-01-4) and *Sodium carboxymethyl cellulose* (CAS Number 9004-32-4) examples of a shear-thinning fluid. This choice of fluid enables the investigation of laminar and turbulent flow as well as the effectiveness of passive AE on complex fluids. Further pressure drop Δp_v is determined in each experiment, showing that different obstructions pump settings and test fluids lead to different readings.

The test rig consists of a water recirculation system powered by an I KA-5 132SSS1 centrifugal pump (Alfa Laval, Sweden). The in-pipe obstructions are all designed in AutoCad 2018 (Autodesk Inc., USA) and extruded by a FlashForge Dreamer 3D printer (Zhejiang Flashforge 3D Technology Co., Ltd., China). Obstruction types can be slotted into the pipe/ 2D PIV chamber and are all significantly geometrically different. Geometries are a cone, cross (four wedges meeting in the central focal point), three triangular aligned holes and a semicircle. The TSI 2D Particle Image Velocimetry system (TSI Inc, USA) uses a green 532 nm Nd-Yag laser (Litron Nano PIV) pulsing at 7 Hz and is synchronised to a single TSI Power

View 4 megapixels (2048 x 2048 pixels) 12-bit CCD camera. The CCD camera is connected to another synchroniser (TSI 610035) which is attached to a desktop PC. The 2D PIV system is controlled by TSI Insight 4G software. For cases in turbulent flow 500 images are taken, whereas for cases of laminar flow 100 pictures. The area is set to 32 x 32 pixels. The pictures are afterwards combined and averaged and debugged in order to determine the flow field. The final pictures are visualised with the help of Tecplot 360 software (Constellation Software, Canada).

Acoustic emission signals are captured with a piezoelectric VS375-M sensor (Vallen Systeme GmbH, Germany) that is linked to a 2.5 kHz to 2.4 MHz (10 Vpp) AEP5H preamplifier (Vallen Systeme GmbH, Germany) along with a DCPL2 decoupling unit (Vallen Systeme GmbH, Germany), a PicoScope 5000 Series oscilloscope (Pico Technology Ltd, UK) and a personal computer using PicoScope version 6.13.15 software (Pico Technology Ltd, UK). 100 buffer, each of a length of 500 ms, a resolution of 16-bit and an amplitude of maximum ±1 V. The sampling number set to 600 kS to ensure that the sampling frequency is at least twice the number of the resonance frequency [16]. The choice of 500 ms is justified as a 10 s buffer is divided that long until the lowest recording time reached where *Fast Fourier Transform* (FFT) spectra are still visually identical.

The recorded acoustic emissions are pre-processed aiming to reduce the size of input data to a level that it can be used for supervised machine learning (ML) on the given computational power of a 8 GB memory. In a first step, the noise must be removed. As environmental noise those frequencies below 4 kHz are considered and the FFT spectrum is extracted. In addition, positive and negative infinitive values are filtered and replaced by the value ±1. The last reduction step is the selection of only the 5,000 FFT values with the largest relative variance.

To make the data available for the Matlab R2018a (MathWorks Inc, USA) Classification Learner Application (CLA) feature scaling and mean normalisation was applied to the reduced FFT spectrum. The necessity of this operation is to have features with a comparable range of values.

Principal Component Analysis (15 components) is enabled for the CLA app. The goal of the PCA is to project the data points of the matrix into an n-dimensional subspace in such a way that as little information as possible is lost and existing redundancy is summarised in the form of correlation in the data points. This enhances the supervised ML performance, contributing to a better support vector delimitation. The dataset for the Machine Learning is further split into in a training (60%), optimisation (20%) and a test dataset (20%). The test dataset is given to the supervised classifiers (e.g. KNN, Decision Tree, SVM) whilst the second dataset is not fed into the CLA to evaluate the accuracy of the algorithms.

For the acoustic data, higher flow rates lead to higher peak values and different pipe obstacles to different FFT spectra. Also, different fluids lead to different amplitudes in acoustic signals with highest peaks for water. Highest accurate predictions in supervised ML are attained with quadratic SVM with accuracy levels being in the region of 95%. This applied as well as to different flow rates, fluids and obstruction types. In addition, 2D PIV supports differences as different types of obstacles lead to different flow regimes and increases in flow rate lead to increases in velocity magnitude. Besides, different fluid types show different flow fields.

Summarising, the application of a single passive acoustic emission sensor linked to supervised ML is proven to give accurate predictions on flow rates, obstructions and fluids. The technology works for Newtonian and Complex Fluids. This may help in the future to predict flow regimes, improve process design and operation conditions. More importantly, the system for single-phased fluid flow, which is not documented in current literature.

Future research can focus on refining the data processing or making use of neural networks for broaden application fields. Also, for each case single algorithms are created. To enhance user friendliness those need tying together or the creation of some sort of cascade algorithm and the creation of a user-friendly interface.

Acknowledgement

The authors wish to express their gratitude for the support received from Philip W Harris and Robert W Sharpe in the engineering workshop at the University of Birmingham. Further thanks goes to Dr Thomas Mills providing necessary training and enabling access to the 3D printer system.

Funding Note

The School of Chemical Engineering of the University of Birmingham supported this work.

The authors declare no conflict of interest.

Funding Note

> **Please note: A reference list will exceed the total word count. Hence, please advise how this can be supplied as well. Thanks. Daniel Hefft**

561329 Event-Based Optimisation for Harvesting Scheduling Under Precision Agriculture

Building on digital technologies (Internet of Things (IoT), big data), Agriculture 4.0 has the potential to enhance precision farming and improve farm system responsiveness and resilience. As part of IoT driven precision solutions, the increasing smart agricultural machinery is expected to catalyse the agriculture sector transformation by shifting the isolated farms towards performance-optimised and vehicle-interconnected agro-ecosystems. Despite the importance, development and application of computer-aided decision-making tool to optimise the precision operations by interconnected smart machinery remains largely unexplored. This study presents optimisation research under the Agriculture 4.0 vision with a particular focus of harvesting operation for a perennial crop, sugarcane.

Sugarcane is one of the most important commodities globally[1]; its production increased fourfold since 1965, achieving 1842 million metric tons in 2017[2]. Notably, BRICS countries representing the growing economy plays significant roles in global sugarcane market i.e. South Africa, Brazil, India and China. Sugarcane represents a highly perishable agro-product due to cane quality deterioration which can be affected by variety, harvesting, and pre- or post-harvest operations[3], thus increase the system complexity involving coordinated planning and scheduling[4]. Thereby, decision-making support to optimize operational planning and to estimate the harvest process quality will be of benefit to the agricultural sectors.

Almost all the mathematical models developed for the planning and scheduling of sugarcane harvesting are based on discretised time interval, which only provides a reference grid of time for all operations (e.g. harvesting, relocation) competing for shared resources (i.e. harvesters). For harvesting operations, the varying geometries of different fields, lead to variation in harvesting time; however, the discrete-time approaches can only provide approximate descriptions of the actual process, which can deviate substantially from the true solutions [5]. The real-time decision making and accurate optimisation solutions over continuous time horizon are of particular importance for coordinated operations and interconnected smart machineries with real-time communication (Agriculture 4.0 vision); this research challenge remains largely unexplored.

In this study we propose an event-based continuous-time formulation to optimise short-term sugarcane harvesting operations, accounting for the interconnected agro-system performances. To demonstrate the model functionality, we present a hypothetical case study based on sugarcane operation using smart harvester at Kwazulu Natal region in South Africa. KwaZulu Natal is a typical sugarcane farm region with an annual water precipitation of higher than 200 mm (historical statistics), which offers desirable climatic conditions for non-irrigated sugarcane growth. The model is formulated in Python 3.7 using Pyomo 5.6.2. In this study, we present a heuristics-based hybrid approach to solve the sugarcane harvesting scheduling problems, which significantly reduced computational time to enable real-time decision-making. Our research highlights the continuous-time algorithm development and applications in precision agriculture to provide accurate optimisation solutions for real-time scheduling of field operations, which are operated in a coordinated manner through smart machineries.

REFERENCES

1. Jena, S.D. and M. Poggi, *Harvest planning in the Brazilian sugar cane industry via mixed integer programming.* European Journal of Operational Research, 2013. **230**(2): p. 374-384.
2. Statistica, *World sugar cane production from 1965 to 2017 (in million metric tons).* 2018.
3. Solomon, S., *Post-harvest deterioration of sugarcane.* Sugar Tech, 2009. **11**(2): p. 109-123.
4. Kusumastuti, R.D., D.P.v. Donk, and R. Teunter, *Crop-related harvesting and processing planning: a review.* International Journal of Production Economics, 2016. **174**: p. 76-92.
5. Floudas, C.A. and X. Lin, *Continuous-time versus discrete-time approaches for scheduling of chemical processes: a review.* Computers & Chemical Engineering, 2004. **28**(11): p. 2109-2129.

576573 Antimicrobial Mesh Materials Applied to Milk and Brine

Pathogenic bacteria are ubiquitous in dairy systems. They are mostly eliminated by pasteurization, though recontamination can occur during downstream processing. Spiral wound membranes are widely used during milk production, and offer an excellent opportunity for mitigation of pathogens. Specifically, the feed spacer in spiral elements has been functionalized to provide antimicrobial activity by adsorbing or contact killing bacteria. The purpose of this research is to develop these materials with applications in milk and brine systems.

A modified polypropylene (PP) feed spacer was produced to control bacterial growth. Vinylbenzyl chloride was reacted with dimethylhexadecylamine to produce an aminated monomer. The resultant product was polymerized by UV polymerization and grafted to the surface of PP. The modified mesh was characterized by mass change and dye testing, followed by Raman and X-ray photoelectron spectroscopies. We have demonstrated antibacterial effectiveness of the polymer against Gram-positive bacteria. Also, scanning electron microscope images before and after the biotest showed a high adsorption of bacteria to the surface. We will discuss our results so far, as well antimicrobial activity of feed spacer in dairy environments.

568142 Investigation of Pipe Cleaning Process Using Planar Laser Induced Fluorescence

Investigation of pipe cleaning process using Planar Laser Induced Fluorescence

Min Zhang[1], Federico Alberini[1*], Kylee R. Goode[1], Katharina Roettger[2], Peter J. Fryer[1]

[1] *School of Chemical Engineering, University of Birmingham, Birmingham, B15 2TT, UK*

[2] *CPI, Wilton Centre, Wilton, Redcar TS10 4RF, UK*

Removal of fouling is a big challenge in food and fast moving consumer goods (FMCG) industries. Understanding of the removal process will benefit product recovery and minimise the waste and energy required. We have used Planar Laser Induced Fluorescence (PLIF) to investigate the fluid mechanically driven cleaning process of a 25.4 mm (*i.e.*, 1 inch) diameter (*D*) Perspex pipe. Configurations with a pipe length (*L*) of 420 or 780 mm were investigated, giving a corresponding *L*/*D* ratio of 16.5 and 30.7, respectively. Cabopol 940 suspensions of different concentrations with a rheology behaviour fitting well to a Herschel-Bulkley model have been used to mimic model food and FMCG products, and water has been used as a rinsing fluid.

The Herschel-Bulkley model is commonly written as Equation 1 (Steffe, 1996).

$$\tau = \tau_y + k\dot{\gamma}^n \qquad (1)$$

where t is the shear stress, t_y is the yield stress, $\dot{\gamma}$ is the shear rate, k is the consistency index and n is the flow index.

The yield tresses, t_y, of the suspensions used in this study ranged between 16 and 52 Pa. All Carbopol suspensions were doped with Rhodamine 6G with a concentration of about 0.45 mg/L. Water was used as a rinsing fluid for removal of the Carbopol suspension from filled pipework. Such configurations allowed illumination of the cross-section of the pipe using a laser sheet perpendicular to the pipe flow. Experiments were performed at water flushing rates ranging from 790 to 1895 L/h. The experiment was conducted by filling the pipe with a suspension, then water was flushed through at a given volumetric flowrate. Meanwhile, a cross-section of the fully filled pipe was illuminated using laser sheet and a camera used to record the entire cleaning process. The system obtained 600-5000 images at a frequency of 25 Hz. Using the acquired image and self-developed image analysis MATLAB script, detailed information of the cleaning dynamics were acquired.

PLIF images show clear fluid breakthrough pattern and different characteristics in removal of different products at pipe cross-section, *e.g.*, zero-yield material flows out easily with water, whereas patch cleaning is clearly observed with yield materials. The cleaning efficiency is quantified by (A_c/A_0)%, where A_c is the area where the suspension has been removed/cleaned up and A_0 is the initial pipe cross-section area fully filled with suspension. Cleaning time for each test is determined from the PLIF images when cleaning reaches 90%. Experiment data are further analysed following Palabiyik *et al.* (2018), by which they obtained a linear relationship between the dimensionless cleaning time (Θ) and a ratio of material yield stress (t_y) to wall shear stress (t_w). The dimensionless cleaning time and wall shear stress could be estimated following Equation 2 and 3, respectively.

$$\theta = \frac{t_c}{t_R}\frac{L}{D} = t_c\frac{U}{L}\frac{L}{D} \quad (2)$$

where t_c is cleaning time obtained from experiment, t_R is the mean water residence time in the pipe, U is the water velocity, L is the pipe length and D is the pipe inner diameter.

$$\tau_w = c_f \times \frac{1}{2}\rho U^2 \quad (3)$$

where

$$c_f = 0.079\,\mathrm{Re}^{-0.25} \quad (4)$$

where ρ, U and Re is the rinsing fluid density, velocity and Reynolds number, respectively.

Plots of Θ against t_y/t_w for the two configurations studied are shown in Figure 1. As it can be seen, a linear relationship is obtained for each configuration, however, the correlations for the two configurations do not follow a same relationship, due to changes with pipe length.

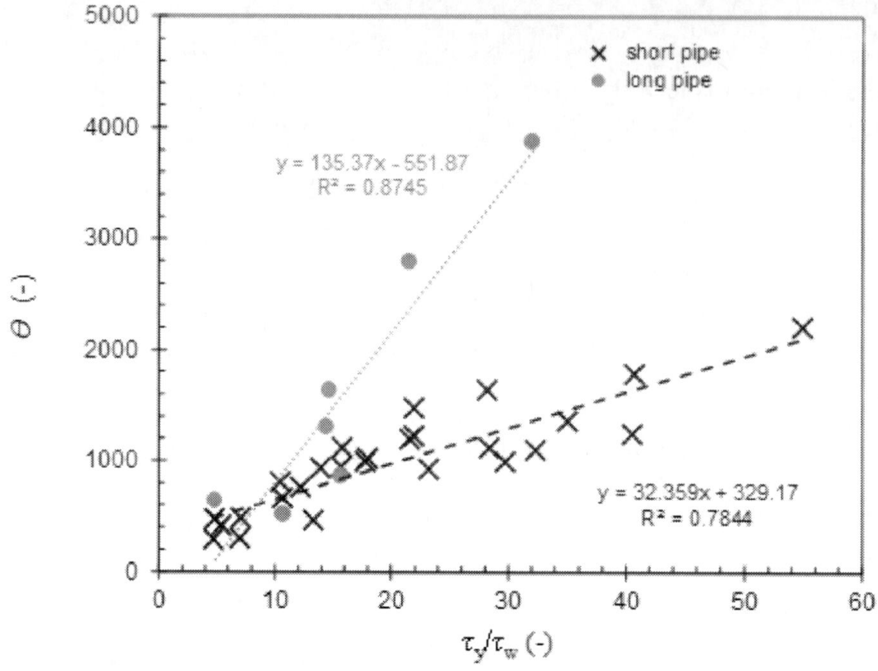

Figure 1 Dimensionless cleaning time (Θ) against a ratio of yield stress to wall shear stress (t_y/t_w) for a short pipe and a long pipe of 420 and 780 mm, respectively. t_y of Carbopol suspension varied from 16 to 52 Pa. Water volumetric flowrate changed from 790 to 1895 L/h.

The results in Figure 1 suggest that the fluid mechanically driven cleaning process investigated in current study is not a simple mass transfer. The above observation extends the approach of Palabiyik *et al.* (2018). Data analysis shows that the cleaning process can divided into stages: (*i*) a big chunk of suspension is removed at the central part of the pipe, which leaves a layer of the suspension attached to the inner pipe wall; (*ii*) the layer at the interface with water is removed gradually, so the thickness reduces with time; (*iii*) a thin layer of the leftover suspension peels off the pipe walls.

Experiment results show that it is feasible to use PLIF to record fluid breakthrough pattern at pipe cross-section and to monitor the entire cleaning process. Data obtained after a detailed PLIF image analysis can give some quantitative information on cleaning efficiency, cleaning time, factors influencing the cleaning efficiency, *etc*. Therefore, PLIF will be used in future work to gain a further understanding of the cleaning mechanism in removal of different types of products from filled pipe.

---

*Corresponding author: Dr Federico Alberini,
Tel.: +441214147397; Email: F.Alberini@bham.ac.uk

References

Palabiyik, I., Lopez-Quiroga, E., Robbins, P. T., Goode, K. R., Fryer P. J., (2018). Removal of Yield-Stress Fluids from Pipework Using Water." Aiche Journal 64(5): 1517-1527.

Steffe, J.F. (1996). Rheological Methods in Food Process Engineering, 2nd edition. Freeman Press, East Lansing, MI, USA.

561018 Model Reduction of Phase-Field Models Describing Crystallisation Phenomena

Crystallisation determines both quality and function of a wide range of food products. For example, texture in ice cream is given by targeting a certain crystal size/distribution that enhances mouth feel, and the first stage of freeze-drying processes – i.e. freezing and thus crystal formation – is key to determine a well-interconnected porous structure that will allow for drying and further rehydration or dissolution of the dried product (e.g. instant coffee). In all these cases, a proper control of the conditions that lead to crystal formation and growth is critical.

Phase Field models have been increasingly used to simulate and predict the formation and evolution of material microstructure and phase change interfacial kinetics. However, these methods usually lead to computationally involved numerical schemes, revealing the need for more efficient computational solutions: i.e. industrial applications, like real-time control and optimisation, require the development and implementation of process models that are capable to operate at faster time scales than the process itself.

This work presents a reduced order Phase-field model for ice crystal formation in food model systems. The model couples heat and mass transfer phenomena, describing the evolution of the solid/liquid interface. We have considered two different model reduction techniques: (i) Proper Orthogonal Decomposition (ii) Laplacian Spectral Decomposition, and we have compared their performance using a range of undercooling and seeding conditions.

Results revealed an overall good reduction performance of both model reduction methods, with savings in computation times up to up to 40% for the POD reduced model and up to 30% for the LSD model. Both reduced models (LSD and POD) exhibited similar predictive capabilities. Simulations showed plate morphologies at high supercooling and more needle-like dendrite formation at low supercooling degrees. In terms of error analysis, RMSE values revealed that the reduced models (both LSD and POD) were able to predict better the temperature field dynamics rather than the more complex ones (i.e. interface kinetics) represented by the phase field.

Overall, this work demonstrates the potential of reduced order approaches for the modelling of phase change processes and also for the development of virtual tools that allow a "fast" (yet accurate) monitoring, design and optimisation of food manufacture operations.

576019 Oleogels As Future *Trans* fat Alternative: A Mathematical Model for an Unsolved Conundrum in Gelation during Scale up Process

Molecular oleogels are supramolecular soft gels, in which vegetable oil is immobilized within the self-assembled network of low molecular weight gelators. Oleogels have found applications in various fields, namely, food, pharmaceutics, cosmetics, petrochemical, tissue engineering and art. Though extensive research has been carried out on oleogels, their industrial usage is still in its infancy. Experimental studies have shown that gelation fail to occur when larger volumes (5ml, 10ml, 15ml and 20ml) were prepared based on the minimum gelator concentration (MGC) that was determined using the smallest oil volume (1ml), a general laboratory scale practice. Gelation failure at larger volumes was evident when tested by two independent studies; vial inversion and rheological studies. This observation was consistent with all the studied molecular gelators, sorbitol dioctanoate (S8), mannitol dioctanoate (M8), and 12-hydroxy stearic acid (12HSA). These results foresee a constraint during fabrication of oleogels at industrial scale; that is to achieve consistency in physical properties during scale up. To understand and predict the gelation behaviour of molecular gelators in vegetable oil of any volume, a mathematical model was developed based on the fact that self-assembly and propagation of gelator networks is governed by rate of cooling. The model indicates that maintenance of minimal thermal gradient from boundaries to center of the vial is necessary to achieve uniform gel propagation across the vial. This provides uniform heat dissipation and gelation times in the vial so that the self-assembled fibers can extend and form network prior to the sedimentation. With these predictions, we hypothesized and confirmed that maintenance of the correct surface area to volume (S/V) ratio which minimizes thermal gradients irrespective of volume of oleogels could result in identical gelation times during scale up. Under this condition, in addition to gelation time, MGC and other physical properties of the gel remained consistent regardless of the sizes of the gel prepared; the other physical properties were melting temperature (T_m), melting enthalpy (ΔH_m), yield stress (Y_s), solid phase content (SPC) and oil binding capacity (OBC).

573800 Spatial Optimization for Land Use Allocation Under Food-Energy-Water Nexus Conciderations

Effective land use allocation is of emergent concern due to land scarcity, diminishing supply of energy and water, and the increasing demand of food globally [1]. Land use patterns are considered as a consequence of competitions between different land use types in a specific land area [2]. The competitions may come from interdependent demand and supply of food, energy and water (FEW), and multiple requirements from social, economic and environmental objectives [3]. The coordination of these competitions and the understanding of the complex nexus formed between FEW components in the land use system is a key challenge to land use spatial optimization, and essential for sustainable development [4].

In this work, a general framework for combined land use optimization and resource management in FEW-N systems is described. An optimization problem is formulated that combines spatial land grids, FEW supply and demand, and multiple objectives. A graph theory based algorithm is employed to facilitate the distribution of land grids. Combined data analytics and mixed-integer nonlinear modeling and optimization methods are used to establish the interdependencies and potentially competing interests among the FEW elements related to different land patterns, along with sustainability, economic benefit, and feedback from multiple stakeholders. A multi-objective optimization strategy is followed for the trade-off analysis empowered by the introduction of composite FEW-N metric as means to facilitate decision-making and compare alternative processes and pathways [5,6]. An existing area in China is selected as a case study to verify the validity of the model. Computational results indicate that the approach can solve land use spatial optimization problems, achieve trade-off solutions for multiple objectives, and coordinate the competitions in the FEW-Nexus.

References

[1] Yuan, K. Y., Lin, Y. C., Chiueh, P. T., & Lo, S. L. (2018). Spatial optimization of the food, energy, and water nexus: A life cycle assessment-based approach. Energy policy, 119, 502-514.

[2] Fotakis, D., & Sidiropoulos, E. (2014). Combined land-use and water allocation planning. Annals of Operations Research, 219(1), 169-185.

[3] Ringler, C., Bhaduri, A., & Lawford, R. (2013). The nexus across water, energy, land and food (WELF): potential for improved resource use efficiency?. Current Opinion in Environmental Sustainability, 5(6), 617-624.

[4] Yao, J., Zhang, X., & Murray, A. T. (2018). Spatial Optimization for Land-use Allocation: Accounting for Sustainability Concerns. International Regional Science Review, 41(6), 579-600.

[5] Avraamidou, S., Milhorn, A., Sarwar, O., & Pistikopoulos, E. N. (2018). Towards a Quantitative Food-Energy-Water Nexus Metric to Facilitate Decision Making in Process Systems: A Case Study on a Dairy Production Plant. In Computer Aided Chemical Engineering (Vol. 43, pp. 391-396). Elsevier.

[6] Nie, Y., Avraamidou, S., Xiao, X., Pistikopoulos, E.N., Li, J., Zeng, Y., Song, F., Yu, J., & Zhu, M. (2019). A Food-Energy-Water Nexus approach for land use optimization. Science of The Total Environment, 659, pp.7-19.

569138 Managing Technological and Ecological Systems in a Watershed While Considering the FEW Nexus, Ecological Carrying Capacity, and the Effects of Climate Change

Accounting for the nexus of food-energy-water (FEW) is essential for sustainable management of watersheds. In addition, since the supply of natural resources is limited, it is also important to consider the carrying capacity of ecosystems that support the activities. For example, thermoelectric power plants, which provide electricity to other activities, are one of the largest contributors to total water consumption and greenhouse gas emissions. However, as the supply of water resources and carbon sequestration ecosystem service from nature is finite in the watershed, the supply of these ecosystem goods and services also needs to be considered for watershed sustainability. Similarly, farming produces food but releases significant amounts of nutrient emissions to the watershed where the supply of water quality regulation service is usually limited. To claim sustainability of a watershed, therefore, interactions between FEW flows and ecosystem flows need to be studied [1].

Considering the impacts of climate change is also crucial for sustainable watershed management. Various climate models (CMs) estimate future climate projections. Those CMs show that we may have an increased risk of water scarcity and warmer water temperature. Climate change could affect the FEW nexus and ecosystem in various ways [2]. For instance, thermoelectric power plants might be displaced by water efficient power plants, such as wind power generation to mitigate climate change. Farming practices may also be affected by climate change due to the change in crop productivity and water availability. Land use and land cover, such as type of vegetation, could be affected as well, and accordingly, causing changes in the supply of various ecosystem services [3].

In this work, we investigate the nexus of FEW and ecosystem capacity in the Muskingum River Watershed in Ohio by considering various technological and agroecological alternatives under current and future climate change scenarios. Technological alternatives include different non-renewable and renewable energy resources, cooling technologies, and CO2 conversion technologies. Technologies for converting CO2 into hydrocarbon products treat CO2 as a valuable resource and help in mitigation net CO2 emissions. Water resources are required to provide hydrogen to the hydrocarbons. Agroecological alternatives include different farming practices and land use changes. Such alternatives not only affect water quality and quantity in the watershed but also affect the provisioning of ecosystem services, such as CO2 sequestration in vegetation and soil. The Soil and Water Assessment Tool (SWAT) is used to explore the agroecological alternatives. The Techno-Ecological Synergy (TES) framework is applied to inspect if any environmental interventions (e.g., CO2 emissions) overshoot their corresponding ecological capacities (e.g., CO2 sequestration) [4].

Some alternatives may be superior in terms of environmental sustainability, but economically expensive. Thus, insights for the sustainable management of watershed are provided by identifying potential trade-offs between environmental and monetary objectives. Monetary values of ecosystem services are estimated to include non-market values of ecosystem services.

The findings of this work will provide insights into alternatives that are 'win-win' in terms of various objectives under several climate change scenarios. Preliminary findings of this study show that dry cooling technologies and no-till farming practice could reduce water quantity consumption and improve water quality while minimizing losses of energy and cost efficiencies. Further, this work could be applied to any watershed to help decision-making of businesses and policymakers to improve watershed sustainability and resilience to climate change.

References

[1] Bakshi, B.R., Gutowski, T.G., & Sekulic, D.P. Claiming Sustainability: Requirements and Challenges, ACS Sustainable Chemistry and Engineering, 6(3), 3632–3639 (2018).

[2] Berardy, A. & Chester, M.V. Climate change vulnerability in the food, energy, and water nexus: concerns for agricultural production in Arizona and its urban export supply. Environ. Res. Lett., 12 (2017).

[3] Schroter, D. et al. Ecosystem Service Supply and Vulnerability to Global Change in Europe. Science, 310, 1333-1337 (2005).

[4] Bakshi, B. R., Ziv, G., & Lepech, M. D. Techno-ecological synergy: A framework for sustainable engineering. Environ. Sci. Technol., 49, 1752–1760 (2015).

565491 Abfews: An Ammonia-Based System for Food-Energy-Water Sustainability

Since the dawn of the 20th century, industrial agriculture and large-scale commercial power generation have revolutionized society, allowing for less expensive and more plentiful food and electricity. However, these industries are problematic with respect to sustainability, consuming considerable amounts of fossil fuels and withdrawing massive amounts of water, much of which is consumed. The water that is returned to the surrounding environment is often done so at conditions which are damaging to local ecosystems. Examples include the elevated temperature of cooling water used in power generation and excess nutrient (e.g. nitrogen, phosphorus) content in agricultural water runoff, especially in areas such as the U.S. Midwest, where tile drainage is used to more quickly remove excess water from crops.

Ammonia lies at the heart of the food-energy-water nexus. Synthetic ammonia is essential for modern society, allowing for higher crop yield per units of land and water. On the other hand, its conventional production is responsible for more than 1% of global greenhouse gas (GHG) emissions and 5% of global natural gas consumption [1]. This has motivated alternative ammonia production concepts. One such concept is to use biomass to replace conventional sources of hydrogen [2]; this is particularly well-suited to an agricultural setting wherein such biomass is naturally present. Another alternative production scheme uses wind- or solar-derived electricity to obtain hydrogen from electrolysis, nitrogen from air separation and subsequently to power an ammonia synthesis process itself [3]. Apart from the ability to reduce the GHG intensity of ammonia production, renewables-to-ammonia has attracted heightened attention due to the potential of ammonia as a means of energy storage to better enable the use of intermittent renewable energy for power generation, or as a carbon neutral combustion fuel. Our recent work explores these possibilities in the context of a system which uses wind energy to make ammonia as both fertilizer and fuel for a local farm, to meet local electrical power demands and to export power to the grid in a predictable and consistent manner [4].

However, the above-described system and others in the literature of a similar type, for example [5] and [6], do not address water sustainability from neither process consumption nor agricultural runoff perspectives. To our knowledge, less research has been dedicated to combined food-energy-water sustainability, with an exception being an integrated local production system which considers crop allocation for food, energy production from crop waste and rain capture for meeting relevant water demands [7]. However, this system does not account for the energy and water aspects of fertilizer and agricultural fuel production, nor does it address the issue of runoff.

In view of the current state of systems engineering research in this area, we in this work propose an ammonia-based system for food energy water sustainability (ABFEWS). This farm-scale system uses renewable energy and agricultural waste to (i) produce ammonia as fertilizer and agricultural fuels, (ii) meet local power demands, (iii) export excess power to the grid in a predictable and consistent manner, and (iv) extract and purify soil water for use in chemical production. The integrated nature of this system offers synergies which aid with both economic viability and sustainability. As with our previous work, time-varying chemical production can act as controllable load to better balance power supply and demand, and ammonia and hydrogen can be used as energy storage media. Additionally, the use of biomass, whether for power generation or ammonia production, reduces electrical power requirements. , Furthermore, purification of soil water simultaneously reduces process water import and tile drainage to surrounding ecosystems.

In addition to the ABFEWS system concept, we present a mixed integer linear programming (MILP) model for its optimal design. For this system, design and scheduling are inextricably coupled due to the hourly and/or seasonal variation in renewable generation, weather phenomena, and demands for power, ammonia and water. Thus, in addition to design decisions (selection and size of units), the model also determines the optimal hourly operating schedule of the system (e.g. production rates, storage levels) over a representative year. We apply this optimization model to a case study based in Morris, MN which considers the use of two 1.65 MW wind turbines and 196,000 kg of corncob biomass to meet ammonia (used as fertilizer and fuel) demand of 40 tons/year and an hourly average electrical power demand of 985 kWh. This case study provides an initial assessment of the economics, GHG emissions reduction, and water conservation afforded by this integrated approach to sustainability, as well as the most promising areas for future research.

References

[1] Swaminathan & Sukalac. Technology transfer and mitigation of climate change: The fertilizer industry perspective. IPCC Expert Meeting on Industrial Technology Development, Transfer and Diffusion, Tokyo, Japan, 2004.

[2] Demirhan, Tso, Powell, & Pistikopoulos. Sustainable ammonia production through process synthesis and global optimization. AIChE J., 2018, DOI: 10.1002/aic.16498.

[3] Sanchez & Martin. Optimal renewable production of ammonia from water and air. J. Clean Prod., 2018, 325-342.

[4] Palys, Kuznetsov, Tallaksen, Reese, & Daoutidis. An ammonia based system for sustainable energy and agriculture: Concept and design optimization. Chem. Eng. Process, 2019, Accepted.

[5] Wang, Mitsos, & Marquardt. Conceptual design of ammonia-based energy storage system: System design and time-invariant performance. AIChE J., 2017, 1620-1637.

[6] Allman, Palys, & Daoutidis. Scheduling-informed optimal design of systems with time-varying operation: A wind-powered ammonia case study. AIChE J., 2018, DOI: 10.1002/aic.16434.

[7] Hang & Martinez-Hernandez. Designing integrated local production systems: A study on the food-energy-water nexus. J. Clean Prod., 2016, 135, 1065-1084.

573018 Comparing the Environmental Implications of National Food Supplies Using Data Envelopment Analysis

The three most fundamental resources to basic human needs are water, energy, and food. As conceptualised by Maslow's Hierarchy of Needs, the first tier that must be primarily realized are basic physiological requirements (food, water, warmth, rest). Until these have been met, individuals are unable to progress up the hierarchy pyramid and fulfil needs of a more complex nature, such as safety and security (Maslow, 1943).

Sustainable use and management of natural resources for clean water, energy and food provisioning is therefore required to ensure the physiological requirements are met for the world's population. The challenges, interdependencies and connections between water, energy and food resource use form the Water-Energy-Food Nexus (WEFN) and achieving WEFN sustainability is inextricably linked to human well-being.

The food production, supply and distribution system has the strongest relationship with human and environmental health, as its sole purpose is to feed and nourish people whilst accounting for 70% of global water withdrawals and 37% of land use. Though the food system's purpose is seemingly simple, current practices and structure has led to over 820 million people facing chronic food deprivation (FAO, 2018) and a rise in diet-related non-communicable diseases, as well as significant contributions to environmental degradation (Tilman & Clark, 2014, Willett et al., 2019).

The existence of two malnourishment extremes highlights the double-ended shortcomings of the current food system. Chronic food deprivation and undernutrition is a symptom of the food system's lack of capacity to ensure universal access and supply of calories. On the other hand, the rise in diet-related non-communicable diseases can be attributed to the failure of today's food system to adequately nourish people with nutrient-dense food supplies (Nelson et al., 2018).

In addition to existing malnutrition challenges, the estimated global population growth to 9 billion by 2050 means the food system must increase its capacity to provide for every person. It must do this whilst also undergoing reform to reduce its contribution to greenhouse gas emissions, freshwater withdrawal demand, soil degradation, land use change, and water eutrophication and acidification. One key strategy is through a shift towards diets consisting of foods with lower environmental footprints (Davis et al., 2016, Bajželj et al., 2014).

In order to facilitate widespread dietary shifts, that alleviate malnourishment and environmental degradation problems, an improved understanding of food supply patterns across countries and regions is crucial. Analysis of global food supply can be combined with an assessment of the food system's performance at providing calories and nutrition at current levels of environmental impact. A performance indicator suitable for this assessment can take the form of a benefit/cost efficiency, with cost being the environmental footprint and benefit as number of calories or nutritional quality.

The environmental footprint of food production and distribution covers a range of indicators that includes land use, greenhouse gas emissions, acidification, eutrophication and freshwater withdrawals. Multiple 'costs' therefore introduce a challenge in calculating a single performance indicator (Chaudhary et al., 2018, Gustafson et al., 2016). This challenge can be overcome through the application of Data Envelopment Analysis (DEA), a non-parametric method to calculate relative efficiency considering entities that manage multiple inputs and outputs, formulated as a linear programming optimisation problem (Charnes et al., 1986).

In the DEA model formulation, a country is considered as a 'decision-making unit'. The country's food supply per capita, combined with different food's environmental impacts (Poore and Nemecek, 2018), is used to estimate the environmental 'costs' (inputs) required to produce and supply that particular composition of food. The five environmental 'costs' (land use, greenhouse gas emissions, acidification, eutrophication and freshwater withdrawals) are considered as one input through a weighted sum. The assignment of subjective weightings is avoided as weighting values are optimised by the DEA model.

The number of calories and nutritional quality associated with each country's food supply per capita are considered as outputs. Nutritional quality is based on the food supply's associated Qualifying Nutrient Balance Score (NBS) (Fern et al., 2015) using estimated supply of nutrients per capita (Smith et al., 2016). The NBS represents the extent to which the per capita food supply can satisfy the daily requirements for a set of nutrients. This includes micronutrients that are of widespread public health concern due to prevalent deficiencies.

In order to address the two extremes of malnourishment caused by current food supplies, this study undertakes two parallel DEA assessments that differ by the type of efficiency it calculates. One type of efficiency measures how a country's food supply makes efficient use of environmental resource inputs to supply calories, and the other to supply nutritional quality (embodied by the NBS). Two efficiency indicators are generated for 139 countries with a reference year of 2010. Therefore, relative 'efficient' and 'inefficient' countries are identified, along with improvement targets.

In addition to the assessment study using DEA, statistical analysis can be undertaken to identify significant correlations between the inputs, outputs and national income datasets. Insightful trends between calorie supply, NBS and Gross National Income (GNI) per capita across countries can be found by applying Spearman's rank-order correlation.

There exists a strong positive correlation between number of calories supplied per capita and GNI per capita ($\rho = 0.69$), whilst a strong negative correlation is found between NBS of per capita food supply and GNI per capita ($\rho = -0.69$). There is also a strong positive correlation found between environmental impact level per 2000-kcal sample and GNI per capita, namely greenhouse gas emissions, acidification and eutrophication ($\rho = 0.62, 0.68, 0.60$). In addition to variation across national income, clustering analysis is applied to identify groups of countries with similar food supply compositions.

The resultant set of efficient units identified by the two DEA models are different. When countries' efficiencies are assessed using the associated NBS of its food supply, efficient countries are from low and lower-middle income groups. The most inefficient units identified are all high-income countries with food supply compositions attributed to the same cluster. Contrastingly, efficiencies assessed using number of calories supplied result in a larger set of efficient countries that span across all income groups.

This study highlights the complexity of the food system by showing the difference between supplying calories and providing nutrition. Through the novel application of DEA, insight is provided into which countries are of interest in the context of improving environmental and nutritional sustainability of food supplies.

References

Bajželj, B., Richards, K. S., Allwood, J. M., Smith, P., Dennis, J. S., Curmi, E., & Gilligan, C. A. (2014). Importance of food-demand management for climate mitigation. *Nature Climate Change*, *4*(10), 924–929. https://doi.org/10.1038/nclimate2353

Charnes, A., Cooper, W. W., & Thrall, R. M. (1986). Classifying and characterizing efficiencies and inefficiencies in data development analysis. *Operations Research Letters*, *5*(3), 105–110. https://doi.org/10.1016/0167-6377(86)90082-9

Chaudhary, A., Gustafson, D., & Mathys, A. (2018). Multi-indicator sustainability assessment of global food systems. *Nature Communications*, *9*(1), 848. https://doi.org/10.1038/s41467-018-03308-7

Davis, K. F., Gephart, J. A., Emery, K. A., Leach, A. M., Galloway, J. N., & D'Odorico, P. (2016). Meeting future food demand with current agricultural resources. *Global Environmental Change*, *39*, 125–132. https://doi.org/10.1016/J.GLOENVCHA.2016.05.004

Fern, E. B., Watzke, H., Barclay, D. V, Roulin, A., & Drewnowski, A. (2015). The nutrient balance concept: A new quality metric for composite meals and diets. *PLoS ONE*, *10*(7). https://doi.org/10.1371/journal.pone.0130491

Gustafson, D., Gutman, A., Leet, W., Drewnowski, A., Fanzo, J., Ingram, J., ... Ingram, J. (2016). Seven Food System Metrics of Sustainable Nutrition Security. *Sustainability*, *8*(3), 196. https://doi.org/10.3390/su8030196

Maslow, A. H. (1943). A theory of human motivation. Psychological Review, 50(4), 370-396. http://dx.doi.org/10.1037/h0054346

Nelson, G., Bogard, J., Lividini, K., Arsenault, J., Riley, M., Sulser, T. B., ... Rosegrant, M. (2018). Income growth and climate change effects on global nutrition security to mid-century. *Nature Sustainability*, *1*(12), 773–781. https://doi.org/10.1038/s41893-018-0192-z

Poore, J., & Nemecek, T. (2018). Reducing food's environmental impacts through producers and consumers. *Science*, (360), 987–992. Retrieved from http://science.sciencemag.org/

Smith, M. R., Micha, R., Golden, C. D., Mozaffarian, D., Myers, S. S., & Obukhov, A. G. (2016). Global Expanded Nutrient Supply (GENuS) Model: A New Method for Estimating the Global Dietary Supply of Nutrients. *PLoS ONE*, *11*(1), 146976. https://doi.org/10.1371/journal.pone.0146976

Tilman, D., & Clark, M. (2014). Global diets link environmental sustainability and human health. *Nature*, *515*(7528), 518–522. https://doi.org/10.1038/nature13959

Willett, W., Rockström, J., Loken, B., Springmann, M., Lang, T., Vermeulen, S., ... L Murray, C. J. (2019). The Lancet Commissions Food in the Anthropocene: the EAT-Lancet Commission on healthy diets from sustainable food systems. *The Lancet Commissions*. https://doi.org/10.1016/S0140-6736(18)31788-4

574232 Systematic Design and Optimization of Water-Energy Nexus for Sustainable Regional Planning

Satisfaction of fuel, power and water demands has always been a driving force for societal growth, while securing scarce energy and water resources from the environment constitutes a great challenge. Climate change and increase of demands along with the population give rise to non-conventional technologies. For instance, shale gas, seawater desalination, and renewables offer new solutions but are highly interdependent [1]. Designing cost-effective and sustainable regional energy-water networks to satisfy end-use demands, with minimal natural resource utilization, while considering non-conventional generating technologies is still an unresolved question.

In this work, we introduce a novel Water-Energy Nexus (WEN) superstructure for the optimal regional energy-water planning. The superstructure is comprised of energy and water resources, energy sources (acting also as water sinks), water sources (acting also as energy sinks), and energy and water consumers. The sources are the intermediate facilities that receive water and energy from the resources, exchange streams with each other and provide upgraded products to the consumers [2]. The formulation considers, (i) finite resource availabilities, (ii) consumer demands to be satisfied, (iii) conversion, intensity and capacity factors of potential sources, (iv) location-allocation of the sources and, (v) water reclaim from the consumers and reuse by extensive water treatment [3]. Optimizing the superstructure allocates the appropriate energy and water sources (combination of generating technologies, generating capacities, connectivity, location), to minimize the total resource supply, generation, and transportation cost. This problem constitutes a generalized instance of Capacitated Multi-facility Weber Problem (CMWP) [4], which also considers, (i) economic objective, (ii) existence of two products (energy and water), and (iii) flows and distances between non-fixed energy and water sources.

The problem is formulated as a mixed-integer nonlinear program (MINLP). The existence of nonlinearities arises from the Euclidean distance definitions. To address the nonconvexity and the poor solution scaling with increasing number of entities (resources, consumers, potential sources), we develop a bilevel decomposition algorithm as an extension from Lara et al. (2018) [5]. The 2D location space is discretized, yielding a lower bounding MILP master problem. Then, by fixing the binary variables of selection of sources, streams, and discretized cells, the original problem is reduced to an NLP subproblem of reduced dimensionality with tighter bounds for distance and coordinate variables. An additional challenge exists due to the stream exchange between non-fixed intermediate sources, which makes the MILP problem of even larger size, especially with refined partitioning. The performance of the algorithm is demonstrated in example problems, and the applicability of the WEN superstructure is demonstrated in regional case studies of Texas Counties of different size.

References:

[1] Garcia, D. J., & You, F. (2016). The water-energy-food nexus and process systems engineering: a new focus. Computers & Chemical Engineering, 91, 49-67.

[2] Tsolas, S. D., Karim, M. N., & Hasan, M. F. (2018). Optimization of water-energy nexus: A network representation-based graphical approach. *Applied energy*, *224*, 230-250.

[3] Tsolas, S. D., Karim, M. N., & Hasan, M. F. (2018). Systematic Design, Analysis and Optimization of Water-Energy Nexus. *Foundations of Computer-Aided Process Design 2019, Paper ID 101*

[4] Brimberg, J., Hansen, P., Mladonovic, N., & Salhi, S. (2008). A survey of solution methods for the continuous location allocation problem. *International Journal of Operations Research*, *5*(1), 1-12.

[5] Lara, C. L., Trespalacios, F., & Grossmann, I. E. (2018). Global optimization algorithm for capacitated multi-facility continuous location-allocation problems. *Journal of Global Optimization*, 1-19.

566645 A Computational Platform for Assessing Organic Waste Management Strategies in the Food-Energy-Water Nexus

The areas of food, energy, and water are crucial sectors in human life, and they are posing significant challenges in economic, environmental, and social areas as the population keeps growing. One of the main challenges is to properly manage the organic waste generated from those sectors to support the well-functioning of the whole society [1]. However, the nature of interactions and interdependencies makes the food, energy and water (FEW) sector a very complex nexus system, and this nature is inherited to the organic waste management problem in the FEW nexus. For example, the dairy industry in the State of Wisconsin contains 1,280,000 dairy cows which can produce dairy food products, but also 6,400,000 kilogallons of manure annually [2]. The cow manure, in turn, can be used in anaerobic digestion to produce biogas and energy. In addition, the organic waste generated in the nexus, such as animal manure and food waste, are highly distributed and usually needs collection and central processing [3]. Considering the complexity of the nexus and the nature of the organic waste, supply chain management (and optimization) has been proven as an efficient method for analyzing such type of holistic, integrated, and multi-media problem and estimating its overall economic and environmental impacts [4].

In the field of process systems engineering, many works have studied the design and optimization of supply chains of organic waste management. A generalized modeling network has been proposed to capture the fundamental elements in the supply chain [5]. In addition, there are numerous case studies which focus on a specific spatial area and contain different types of biomass as basic feedstock, such as cow manure, wastewater sludge, and crop residues [6,7]. These case studies demonstrate how appropriate waste processing can achieve monetary savings and environmental improvement for communities, and how this information is valuable for different stakeholders and policy-makers. However, such types of work highly rely on dependable data inputs, especially in product conversion factors and product conversion costs. And for some cutting-edge technologies (e.g. struvite recovery, co-digestion of different feedstocks), it is difficult to get trustable data.

To provide best practices for integrating computational and experimental work in organic waste and nutrient management science, and to enhance stakeholder engagement, we have developed an easy-to-use computational platform. The platform has two modes: a demo mode and an expert mode. In the demo mode, we have included results and conclusions from several pre-solved cases to allow stakeholders to do parametric studies and observe how individual-level changes (e.g., investment on certain technologies) can impact the whole supply chain system. And in the expert mode, the stakeholders are able to define the scope of their own organic waste management problem using a graphic interface and upload their own data (e.g., product conversion factors for certain technologies). Some basic databases containing data from published literature, the technical support for modeling and solving the supply chain problem will be provided. The results visualization is also automated using a back-end GIS tool. The capability of this platform will be discussed, and a simple tutorial will be given.

References:

[1] Finley J W, Seiber J N. The nexus of food, energy, and water[J]. Journal of agricultural and food chemistry, 2014, 62(27): 6255-6262

[2] USDA National Agricultural Statistic Service: Milk Production and Mild Cows. https://www.nass.usda.gov/Charts_and_Maps/Milk_Production_and_Milk_Cows/

[3] Angelidaki I, Ellegaard L. Codigestion of manure and organic wastes in centralized biogas plants[J]. Applied biochemistry and biotechnology, 2003, 109(1-3): 95-105

[4] Pan S Y, Du M A, Huang I T, et al. Strategies on implementation of waste-to-energy (WTE) supply chain for circular economy system: a review[J]. Journal of Cleaner Production, 2015, 108: 409-421.

[5] Sampat A M, Martin E, Martin M, et al. Optimization formulations for multi-product supply chain

networks[J]. Computers & Chemical Engineering, 2017, 104: 296-310.

[6] Hu Y, Scarborough M, Aguirre-Villegas H, et al. A Supply Chain Framework for the Analysis of the Recovery of Biogas and Fatty Acids from Organic Waste[J]. ACS Sustainable Chemistry & Engineering, 2018, 6(5): 6211-6222.

[7] You F, Wang B. Life cycle optimization of biomass-to-liquid supply chains with distributed–centralized processing networks[J]. Industrial & Engineering Chemistry Research, 2011, 50(17): 10102-10127.

582237 Complexities in Simple Mixing

The mixing in many food systems seems quite simple. Many of the relevant processes are single phase Newtonian fluids, designed for a specific mixing time or residence time distribution. Stirred tanks and stirred tanks in series are common vessels for single-phase reactions such as enzymatic modifications. Even these simple cases can show non-intuitive results. We will discuss three such examples: residence time distributions in an unstirred-tank designed to mix feed streams, residence time in a series of CSTRs approximating a plug-flow reactor, and mixing times in a shear-thinning solution.

561004 Steam Stripping of Aroma from Coffee: Multi Scale Modelling of Extraction Behaviour

Aroma is a major factor in consumer satisfaction in drinking coffee, and more than 1000 compounds have been identified n coffee spanning various chemical groups. During instant coffee manufacture, aroma is extracted from fresh roast and ground coffee, to be added back before the final drying stage, preventing evaporative and thermal losses.

One means of aroma extraction can be steam stripping of a coffee bed, the moisture content of which may range from dry to fully saturated. The product aroma profile is largely determined by this process, therefore, there is a need to understand how key aroma compounds extract upon contact with water and steam at different conditions.

A theoretical model of steam stripping a coffee bed has been constructed, consisting of four stages of mass transfer, at the scale both of the individual particle and the whole bed; oil-to-water aroma release; diffusion through the porous coffee grains; partition into the gas phase; and advection up the packed column. Simulations showed three rate-limiting extraction mechanisms related to oil-water partitioning, diffusivity and water-air partitioning.

Experimental data has provided evidence for these three extraction behaviours but there are compounds limited by other factors not anticipated by the model. These are believed to arise from interactions with the soluble coffee components and the solid coffee matrix, a hypothesis supported by further experiments.

Process variables including the grind (particle) size, moisture level and distribution in a coffee bed have been tested and reveal a range of responses depending on the chemistry of the aroma. Moisture may help mobilize the soluble residues on the surface of the coffee grain which may contain aroma and thus assist its extraction, or it may enable the aforementioned interactions to take place, thus limiting volatization of aroma.

Experimental and modelling results have been combined to reveal a wide range of extraction kinetics of volatile aromas in three phase steam stripping of roast and ground coffee. This knowledge reveals how manipulation of the process conditions - time, moisture, grind size, geometry - can influence select chemical groups, enabling the optimization of the process in tailoring an instant coffee aroma profile through process engineering.

582249 Quantifying the Extent and Rate of Caking in Food Powders

"Caking" is the phenomenon where materials gain strength during storage due to changes in the particle/particle interactions. A significant number of materials, including raw feeds, additives and process intermediates, from a wide range of industries are susceptible to these changes. Following storage, previously free-flowing and easily processable powders can have limited ability to pass uninterrupted through their process and this can have a detrimental impact on product quality. Therefore, the ability to quantify the extent, strength and rate at which a powder cakes is vital to process operators and designers. This presentation demonstrates how the caking of food powders can be quantified using powder flow characterization techniques.

Caking can occur via a number of different mechanisms: mechanical, thermal, environmental and/or chemical. A combination of mechanisms is discussed in this study, where the high levels of fats and sugars within food powders can also intensify the strength and rate of caking. Some caking mechanisms can be controlled by managing environmental conditions, however this can be impractical and uneconomical. Accurately quantifying unacceptable characteristics and the stimuli that contribute towards caking for particular powders is therefore highly advantageous.

A range of food powders were evaluated to investigate the effect of extended storage time and environmental conditions on their flow properties. Samples were placed into 25 ml cylindrical vessels and initially conditioned using an FT4 Powder Rheometer® (Freeman Technology Ltd, UK). Conditioning was completed by passing a bespoke, twisted blade through the powder along a prescribed helical pathway. This promoted the formation of a uniform powder bed with a repeatable stress state within the sample. The test samples were then stored in an environment at controlled humidity and temperature levels for predetermined time periods. Stored samples were analyzed using the same bespoke blade and test parameters to measure the powder's resistance to motion and quantify the relative change in flow properties as a result of caking.

Analysis of the powder's resistance to motion with respect to blade position provides strong evidence of localized caking within some materials. At elevated humidity levels, some powders gained strength at the powder-air interface, forming a "crust". It was possible to easily evaluate both the strength and depth of the crust and improve understanding of moisture migration through the powder bed.

This study demonstrates that caking is a complex phenomenon and that powder flow characterization is a relevant technique to quantify the progression of caking as a function of time, relative humidity and temperature. In turn, this characterization can assist with adapting the process environment or parameters to limit caking and retain optimal processability.

570446 Characterization of New Hop Varieties and Beers in Support of New Mexico Craft Brewing Industry

In New Mexico, as in many states in the U.S., the craft brewing industry is continuing to expand as new breweries and brewpubs open, and farmers begin to grow hops and malting barley to provide local ingredients. Sustainability of these new enterprises depends on the availability of training, analyses for QA/QC, information about best practices, and networking along the supply chain. To help support these industries, NMSBrew—the brewery engineering education and research program at New Mexico State University—has developed capacity to provide analysis of hops from local hop growers for bittering acids, hop storage index (aging), essential oils, and polyphenols, and of wort and finished beers for local brewers for density, alcohol by volume, and concentrations of carbon dioxide and oxygen. Early work on hop characterization has focused on new varieties of native hops, *Humulus lupulus neomexicanus*, as well as traditional hop varieties grown in the lower latitudes and drier/hotter climate of the desert Southwest. Results indicate that *H.l. neomexicanus* hops contain higher levels of polyphenols, especially xanthohumol, and unique essential oil components.

583934 Computational Fluid Dynamics Modeling of Heterogeneous Red Wine Fermentations

Red wine fermentations are performed in the presence of grape solids (skins and seeds) to ensure extraction phenolics. This results in a heterogeneous fermentation system, as the solids float to the top to form a porous solid-liquid "cap." Modeling of red wine fermentations is, therefore, complex and must consider spatial heterogeneity to predict fermentation kinetics. We have developed a reactor engineering model for red wine fermentations that includes the fundamentals of fermentation kinetics, heat transfer, diffusion, and compressible fluid flow. COMSOL was used to solve all components of the model simultaneously utilizing a Finite Elements Analysis (FEA) approach.

This reactor engineering model was compared against experimental temperature profiles in red wine fermentations, to good effect. The model then permitted the simulation of difficult to measure parameters, such as spatial productivity gradients and fluid velocity profiles. This model was applied to examine the impact of vessel geometry on fermentation dynamics, and finally to investigate nontraditional fermentation vessels. Follow up work (covered in a subsequent talk) extended the reactor engineering model to the prediction of phenolic extraction during red wine fermentations.

583935 Research into the fermentation process at the University of Florida

Fermented beverages have been around for thousands of years and have been utilized by nearly every culture. These products have become one of the latest food trends, including novel products and opportunities for innovation. This presentation will detail some of the projects currently underway at the University of Florida, as well as opportunities and challenges for the industry. One such project is the application of vacuum during industrial fermentations. Changes to industrial brewing fermentation processes are generally rare, largely due to the impact on the viability of organisms involved. This is related to environmental stresses during the process such as: osmotic pressure, ethanol concentration, dissolved CO_2, *etc.* Fermentation under vacuum has been shown to dramatically increase the rate of the fermentation, possibly by removing inhibitor compounds such as CO_2. This study explored the applications of this finding in brewing, high density brewing and distillation industries. Specifically, the effect of vacuum pressure control on ethanol production rates, volatile loss (including ethanol) and flavor development for standard and high gravity brewing operations were examined. This was accomplished by controlling the temperature and pressure in pilot fermentations ranging from 4-40L. "Diamond" Lager and "Belle Saison" Ale yeasts, provided by Lallemand were used for High Gravity (HG) and Very High Gravity (VHG) fermentations respectively. The modified process (3.5psia) was compared to a control process at standard pressure (14.7psia). Volatiles analysis was performed using a Stratum purge and trap unit, and identified using GC-MS. This research showed that at 15°C, a lager fermentation (14-15°Brix) under vacuum pressure fermented faster (~15-30%), with an increase (~15%-100%) in the number of cells in suspension (depending upon the initial density). The rate of sugar consumption under vacuum was greater compared to the control, thus increasing the rate of ethanol production. Vacuum fermentation reached the final batch ethanol formation 80 hours (~3days) before the control. The removal of volatile CO_2 is one hypothesized mechanism for these results. When a condenser was attached to the tank for volatile retention, there was no measured loss of ethanol. Under vacuum, more higher alcohols (~86%) and esters (~78%) were present in the final product compared to the control. This research shows that vacuum pressure has a positive effect upon the fermentation rate. Additionally, vacuum fermentation shows a higher generation of volatile compounds, having a significant impact upon the product flavor profile. This research can also be exploited by industries where the volatile metabolites produced are the desired product.

583936 The incredibly diverse microbiome of the Kombucha SCOBY

Kombucha is an acidic, effervescent, slightly alcoholic beverage, made by fermenting sweetened tea with a complex microbial consortia known as the SCOBY (Symbiotic Culture of Bacteria and Yeast). Originating in the far east and consumed for at least the past 2,000 years, until recently Kombucha has typically been home-brewed at relatively small scale.

Today, Kombucha is part of a consumer mega-trend towards fermented and minimally-processed products, and the Kombucha industry has consequently experienced explosive growth. The problem many producers face is how to scale-up while meeting specifications and tighter production schedules. This problem is exacerbated by a lack of knowledge concerning the microbial composition of the SCOBY, and how this relates to fermentation efficiency and product quality.

To address this knowledge gap, we collaborated with Kombucha Brewers International to survey the microbial complexity of commercial production SCOBY across the United States using microbiome sequencing approaches. Our results show that individual SCOBY contain upwards of a dozen yeast and bacterial genera, highlighting the microbial complexity of Kombucha production. Clustering of the microbial profiles revealed five main SCOBY types discriminated by their dominant yeast and bacterial genera. In laboratory-scale fermentations, representative SCOBY from these groups differed in the rate at which they acidified sweetened tea, and produced kombucha with differing chemical compositions.

These results provide the first large-scale insight into the diversity of Kombucha SCOBY and reveal a level of interchangeability between different microbial genera to achieve similar function. Ongoing research is using this framework to ask how individual yeast and bacterial species contribute to Kombucha style, and to what extent an individual SCOBY reflects the production environment in which it is used.

583937 Vegetable fermentation: What is the starter culture?

Fermentation produces thousands of foods, including pickles, cheese, and even coffee. This microbial modification of food is also a commercially and culturally relevant practice that feeds billions around the world. Despite recent interest in microbial communities of fermented foods, there has been little inquiry into the bacterial community dynamics of vegetable fermentation. Utilizing 16S rRNA amplicon sequencing, we can get a novel perspective on the microbial community of naturally fermented products such as sauerkraut and kimchi. For this work, we have partnered with local industrial producers to track the microbial composition of the product from ingredients to the supermarket shelf. In our recent kimchi work, the producer was interested in introducing a vegan version of the product (kimchi typically contains fermented seafood). To determine whether a substitution of miso for seafood confers a difference in the microbiome of kimchi fermentation, we used next-gen sequencing to analyze the microbiome before, during, and after production. The key finding is that both the vegan and nonvegan community start vastly different in composition and then converge to the same community in the final product. This result indicates that both versions would provide the same "probiotic" benefit. Most interestingly, the bacteria in the factory environment look very similar to the microbiome of the kimchi, possibly indicating that the fermentation environment plays a significant role in the final community. One can think of this as a form of terroir (similar to wine). Interestingly, this was not true of sauerkraut as we found that the environmental bacteria of the production facility were not significantly reflected in the final product, possibly due to the more selective environment of sauerkraut fermentation or the multiple-use nature of the sauerkraut producer.

583938 Apple juice chemistry and cider fermentation: Phenolic and nitrogen composition influence fermentation rate and cider aroma

The phenolic and nitrogen composition of apple cultivars used for cider making are highly variable, with concentration ranges typically exceeding those previously evaluated in beverage fermentation research. Heritage, or traditional, cider apples are often valued for their relatively high concentration of tannins (phenolic compounds) and/or acids, which impart desirable sensory characteristics of tartness, bitterness, and astringency to cider. Most cider fermentation practices do not account for the potential effects of high concentrations of phenolic compounds on fermentation outcomes. The objective of this study was to determine the impact of three phenolic compounds: ferulic acid, *p*-coumaric acid, and chlorogenic acid—at concentrations reported in apples—and their interactive effects with yeast assimilable nitrogen (YAN) on cider fermentation kinetics and cider aroma. High concentrations of ferulic acid (300mg/L) resulted in lower fermentation rate and incomplete fermentation, and these negative effects could be partially remediated through addition of YAN (100mg/L and 200 mg/L), a common cider making intervention. A sensory sorting task showed that changes in pre-fermentation *p*-coumaric acid and ferulic acid concentrations influenced aroma, while chlorogenic acid did not, and that these effects were significantly different among the two yeast strains evaluated (W15 and EC 1118). This work demonstrates the complex and interactive effects of fruit chemistry on cider fermentation outcomes, and cautions against broad extrapolation of beverage fermentation management practices across diverse matrices.

583942 More Sustainable Winemaking

The wine industry is facing urgent challenges to minimize energy and water usage, wine loss, waste generation while working within the constraints of labor availability. The industry has grown by often scaling up batch processes without introduction of benefits enabled by continuous or semi-continuous processes. Two such processes are those for stabilizing wines from precipitating potassium bitartrate and developing protein hazes. The traditional in-tank approaches result in excess usage of water and energy, loss of wine, and generation of solid waste. Two liquid-solid separations processes are being adapted to benefit the wine industry: Fluidized Bed Crystallization for removing excess potassium bitartrate and Ion-Exchange for removing excess proteins. These approaches are based upon similar chemical and physical interactions to those of traditional in-tank treatments. Our results, which build on prior work, continue to demonstrate their efficacy under a variety of wine chemistries and process conditions and aim to provide the industry with another alternative to existing treatments which can minimize energy and water usage, wine loss and waste generation.

553054 Evaluation of Ultrasound By Direct Contact on Convective Drying of Banana (*Musa paradisiaca*) Slices Using Principal Component Analysis and Response Surface Methodology

Conventional convective drying is one of the latest techniques used to reduce the amount of water in foods, however, due to the long times of exposure to high temperatures, these techniques can reduce the sensory quality, in addition to generating high costs associated with consumption of energy. An alternative to improve efficiency in these processes is to use innovative tools such ultrasound (US) application by using direct contact during drying to improve quality and keep nutritional content of the product with reduction in time, cost and temperatures of operation.

In this study, a Box-Behnken experimental design was used to study the influence of three independent variables: US time-frequency (t), US power level (Pot), and air temperature (T) on the drying process time, apparent density, total polyphenol content and color difference as responses of the banana slices. Seventeen experiments were performed which involved 12 three-level experimental points and 5 repetitions in the central point. Fisher's statistical testing was performed for the analysis of variance (ANOVA) for quadratic regression equations. Results of each response variable were analyzed according to the Response Surface Methodology, analyzing tendencies and regions of operation with contour graphs. In addition, models were proposed for each response variable, which included the input factors of the experimental design proposed. These models were reduced in order to eliminate non-significant terms and were carried out validation of the statistical models proposed. After performing the reduction of non-significant terms, the ANOVA analysis showed a good fit for most of the response variables. PCA was performed to evaluate effect of ultrasound on quality characteristic of banana-dried slices. The PCA analysis showed that the drying time and the apparent density of the samples was strongly related to the temperature of the process. Moreover, power level of the ultrasound influenced the color variation of the dry sample respecting fresh fruit. Finally, US time-frequency affected total polyphenol content of dried samples. Analysis of both tools like PCA and RSM allowed knowing the effect of the independent variables on the drying operation and quality parameters on banana dried slices.

561020 Energy Usage in Food Manufacture

The food industry, the largest manufacture sector globally and also one of the major contributors to GHG emissions, is facing significant environmental challenges that are building up pressure on food security. As the global population keeps growing, water, food and energy demand are expected to increase too - up to 50 % by 2050. A starting point to reduce such demand is to identify energy consumption hot spots in food manufacture operations. In this context, this work provides a comprehensive and updated database for energy processing usage – within the UK and globally. Specific energy consumption (SEC) data for products, processes and food distribution were collected from 44 literature sources based on product-based energy intensity metrics. To account for changes in technologies, processing and fuel efficiencies, and structural changes within sectors data was sorted in three groups: prior to 2000, 2000 - 2004, and 2005 – 2015.

Findings show that thermal processes consume most of the total processing energy – generated extensively from fossil fuels (e.g. > 65% average of the dairy sector), which shows potential for alternative technologies to be implemented. Emerging technologies are typically electric powered, so the energy could be supplied from renewable resources. Data also revealed that the most energy intensive food products are powders (instant coffee and milk powder), fried goods (French fries and crisps) and bread, all them products that involve phase-change in their manufacturing, such freeze-drying or drying. The ways in which many food processes are carried out require first the addition of water to the product followed by its removal (either by evaporation or sublimation), revealing water flow streams that can be optimised and recycled. In the meat and dairy processing sectors, energy and water use have increased due to a rise in hygienic standards and cleaning requirements, which represents an opportunity for hygienic design developments. The challenge here is to introduce energy and water-usage saving policies into the manufacturing protocols without either affecting the quality of the foodstuff or making it less safe. Developing food formulation with lower moisture contents could be a solution.

Energy use in the UK follows the same trends, although a lack of studies that assessed food products playing a central role to the UK – e.g. ready to eat meals, sandwiches – was also identified. An analysis of the relationship between energy density (calories intake), weekly consumption and processing energy use in the UK revealed that the some of the more demanded processed, nutrient-dense foods (e.g. cakes, cookies) require also higher amounts of energy during processing.

In terms of food distribution, a shift towards distributed supply systems – as opposed to centralised ones – and to local production has been identified. For example, more than 98% of all foods within the UK are transported by road, with increasing distances travelled in recent years. However, the environmental benefits of these changes are not always clear, which reveals the importance of global environmental assessment tools (such LCA, carbon and water footprint) that take the whole food chain into consideration.

Overall, this work reveals a series of needs and points towards current trends in food manufacturing that can help both stakeholders and policy makers to allocate resources more effectively and design new food manufacturing protocols that increase sustainability and security of food chains.

583939 Sustainability of Nano-Packaging Use for Food Shelf Life Extension: Life Cycle Assessment Perspective

Food waste is a severe economic and environmental problem in industrialized countries. Nano-packaging has recently become a promising technology for food preservation with increased shelf life, and hence has the potential to reduce food waste. However, the production of nanomaterials can generate a variety of environmental burdens, including the use of energy, water, and other resources, and the associated emissions. In this study, a cradle-to-grave life cycle assessment was conducted to evaluate the net carbon footprint of four nano-packaging systems containing selected food products, considering the direct impact of nanomaterial production and the indirect impact caused by food waste reduction. With the functional unit defined as 1 kg of food product and the required amount of nano-packaging materials, the global warming potential (GWP) associated with the nano-packaging system was highly affected by the concentration of incorporated nanomaterials and the type of packaged food. Furthermore, the packaging-food ratio played an important role in the relative contributions of nanomaterial, polymer, and food to the GWP of a nano-packaging system. A consumer behavior study was also performed to indirectly quantify the amount of food waste reduced resulting from the shelf life extension of food products. The results showed that the GWP decreased due to food waste reduction was significantly higher than the additional GWP produced by nanomaterials, indicating a net beneficial impact of the use of nano-packaging on the carbon footprint of food preservation. This proof-of-concept study presents a novel methodological framework to evaluate the environmental trade-offs of nano-packaging for food preservation, which can help the public better understand the net carbon footprint of nano-packaging from a life cycle perspective. The results of this study are expected to provide the groundwork for food manufacturers to make more informative decisions on the applications of this emerging technology in food industry.

583940 Water Sustainability: The Coca-Cola Company's Story - From Risks to Opportunity

As early as 2002 The Coca-Cola Company faced risks to its business based on water sustainability. In 2003 the Company examined various types of water risks, conducted a risk survey of manufacturing sites, and determined its risk profile and a series of mitigation strategies. Subsequent water risk assessments in 2008 and 2015 have formed the basis for additional mitigation strategies and helped establish the path forward for the Company.

AUTHOR INDEX

AUTHOR INDEX